PE PREPARED

CIVIL PE PRACTICE EXAM

BREADTH EXAM

VERSION B

Congratulations on your decision to take the Principles and Practice of Engineering Exam for Civil Engineering!

PE Prepared was created by real, practicing civil engineers to give E.I.T.s and E.I.s like yourself a leg up on test day. We strove to author realistic questions at the right level of difficulty, with detailed, step-by-step solutions to help you learn the content that is going to be on the exam.

Our questions aren't harder than they need to be, but aren't easier either. They should take less than 6 minutes to solve. Take PE Prepared practice exams as a realistic simulation of exam day to measure your level of preparedness, or simply use them as a bank of practice questions while you study. The choice is yours!

Remember: Civil engineering is a noble profession. Civil engineers make the world a better, safer, and healthier place for people to live in. Congratulations again on your decision to take the PE exam, you're going to pass!

Check out our website at PEprepared.com for study tips and free resources for the exam. We can be contacted at ask@PEprepared.com.

START

CIVIL PE PRACTICE EXAM

BREADTH EXAM

VERSION B

101. A wall plan is shown below. The area of formwork required to form the cast-in-place concrete walls for the structure is most nearly (neglect overlapping at corners):

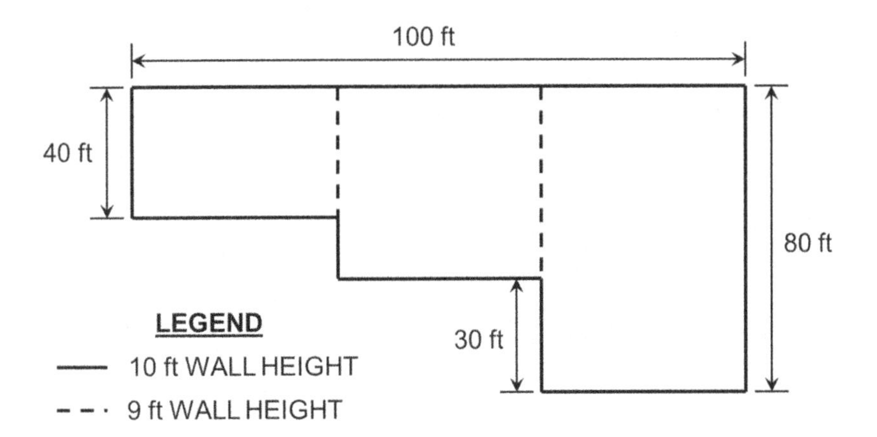

(A) 1,620 SF
(B) 4,410 SF
(C) 7,200 SF
(D) 8,820 SF

102. Which of the statements below is most correct for a project schedule where a critical path exists?

(A) Total float and free float are equivalent.
(B) Free float is the duration an activity can be delayed without violating a set schedule constraint or extending the project completion date.
(C) Free float is the duration an activity can be delayed without delaying the early start date of immediately subsequent activities.
(D) Total float is the duration between the early finish date and the late finish date for the final activity in the schedule.

103. A state transportation agency needs to construct a four-lane roadway (two lanes in each direction) for a length of one mile. Each lane will be 15 feet wide, and there will be an 8-foot wide shoulder on each side of the roadway. If hot-mix asphalt is used to pave the roadway, an 8-inch thick asphalt section is required. If reinforced concrete is used to pave the roadway, a 6-inch thick concrete section is required. The cost of hot-mix asphalt is $100/ton. The cost of reinforced concrete is $270/yd^3. Assume an asphalt density of 145 lb/ft^3. If the most cost-effective option is selected, the cost of paving the roadway is most nearly:

(A) $1,780,240
(B) $1,800,650
(C) $1,939,520
(D) $2,006,400

104. Examine the project schedule below. All relationships are start-finish unless shown otherwise (SS= start-start, FF= finish-finish). Durations are shown in days. Which path is the critical path?

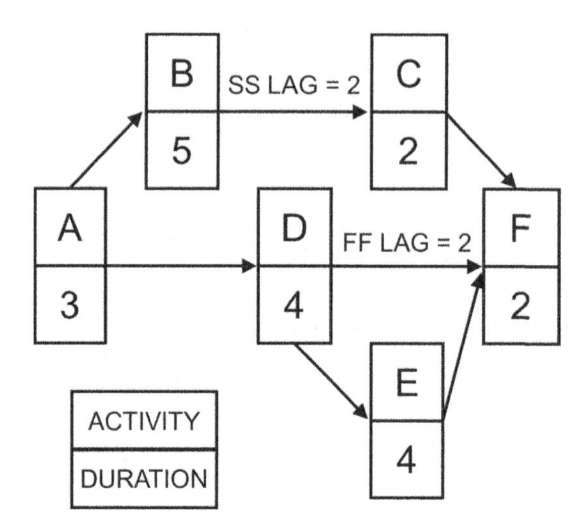

(A) A-B-C-F
(B) A-D-F
(C) A-D-C-F
(D) A-D-E-F

105. A contractor is preparing a project schedule for the construction of a cast-in-place retaining wall. The placement and compaction of crushed aggregate for the footing base is estimated to take 4 days. The preparation of formwork can begin immediately after the base course is placed and compacted. Formwork preparation is estimated to take 5 days. The concrete pour is estimated to take 1 day, and the concrete pour can take place immediately after the formwork is prepared. Placement and compaction of backfill behind the wall cannot begin until the concrete has cured for 10 days. The estimated duration to place and compact backfill is 4 days. The amount of float for the backfill placement and compaction operation is most nearly:

(A) 0 days
(B) 4 days
(C) 10 days
(D) 14 days

106. A contractor is planning to install a new storm drain manhole. Given the following data, the excavation depth to the required subgrade elevation is most nearly:

Feature	Elevation
Existing Grade	120.85'
Finished Grade	122.45'
Manhole Rim	122.45'
Outlet Pipe Invert	110.65'

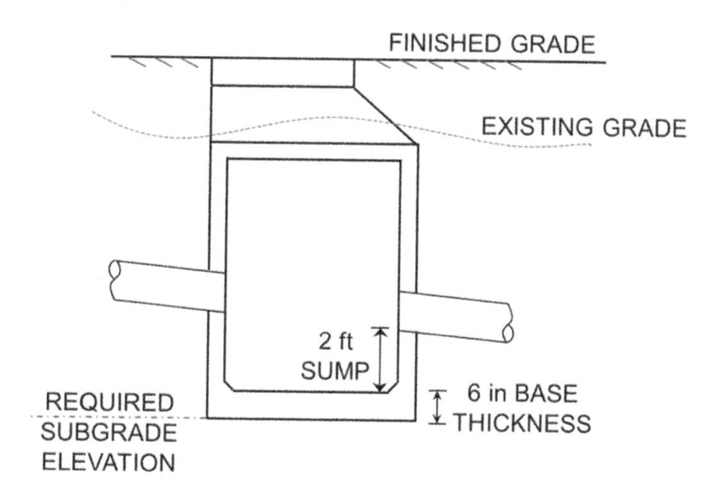

(A) 10.2 feet
(B) 12.2 feet
(C) 12.7 feet
(D) 14.7 feet

107. A long-reach excavator boom is required to construct the excavation shown below. The long reach boom is comprised of two arms, Arm A and Arm B. Several boom configurations are available, but the length of Arm B is equal to 1.25 times the length of Arm A for all configurations. Arm A can only deflect to the horizontal position, as shown below. The total required length of the boom (Arm A plus Arm B) is most nearly:

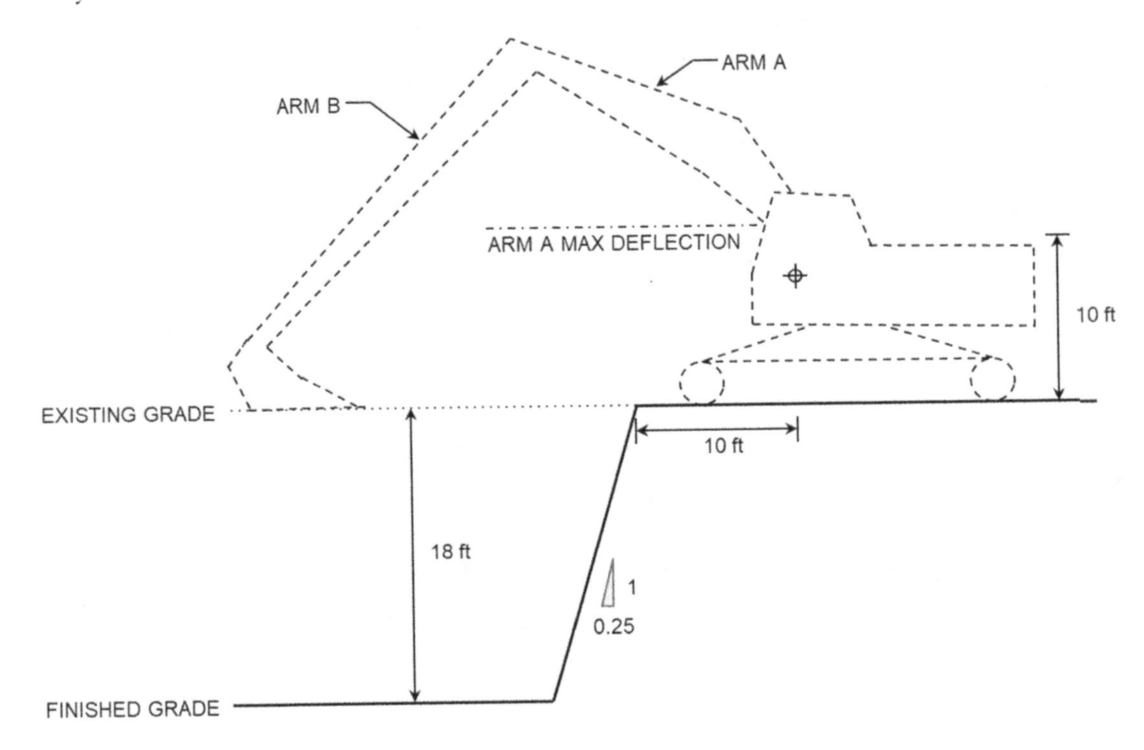

(A) 14.5 ft
(B) 28.0 ft
(C) 32.6 ft
(D) 50.4 ft

108. The figure below shows two lateral earth pressure resultant forces on a reinforced concrete cantilevered retaining wall. In order to calculate the factor of safety against sliding and overturning, which statement is the most correct?

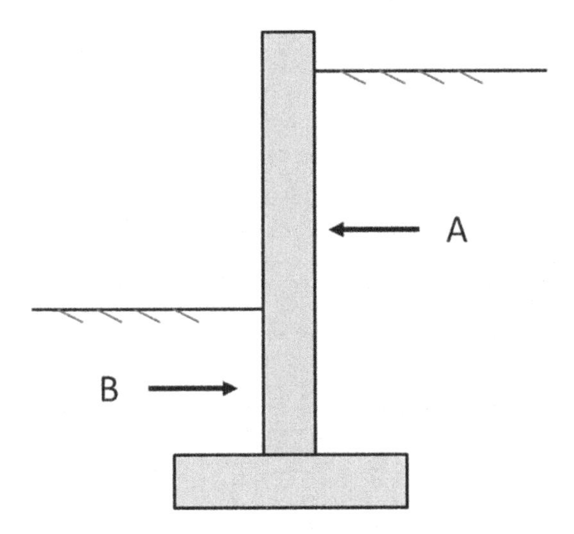

(A) Resultant "A" is the active earth pressure resultant and Resultant "B" is the passive earth pressure resultant.
(B) Resultants "A" and "B" are active earth pressure resultants.
(C) Resultant "A" is the passive earth pressure resultant and Resultant "B" is the active earth pressure resultant.
(D) Resultants "A" and "B" are passive earth pressure resultants.

109. The allowable bearing pressure on a sandy gravel subgrade is determined to be 0.75 tons/ft^2. The column shown below is required to support a live load of 2.25 kips and a dead load of 6 kips. The minimum dimension "B" of the shallow, square footing is most nearly:

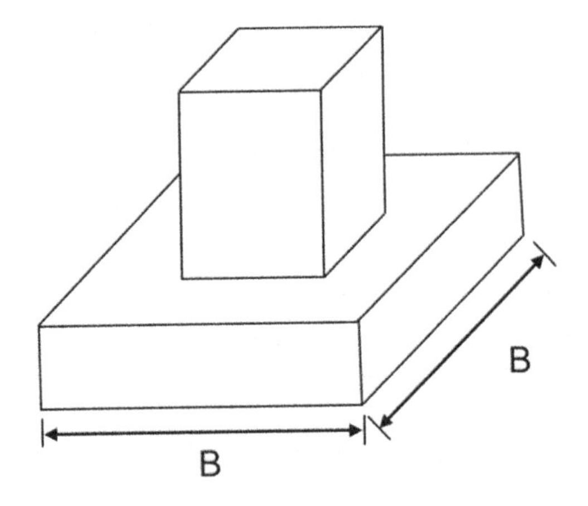

(A) 2' - 5"
(B) 3' - 2"
(C) 4' - 0"
(D) 5' - 6"

110. A structure is constructed on the soil column shown below. The net increase in vertical stress is 200 psf at the center of the clay layer. The clay is normally consolidated. The settlement resulting from primary consolidation of the clay layer is most nearly:

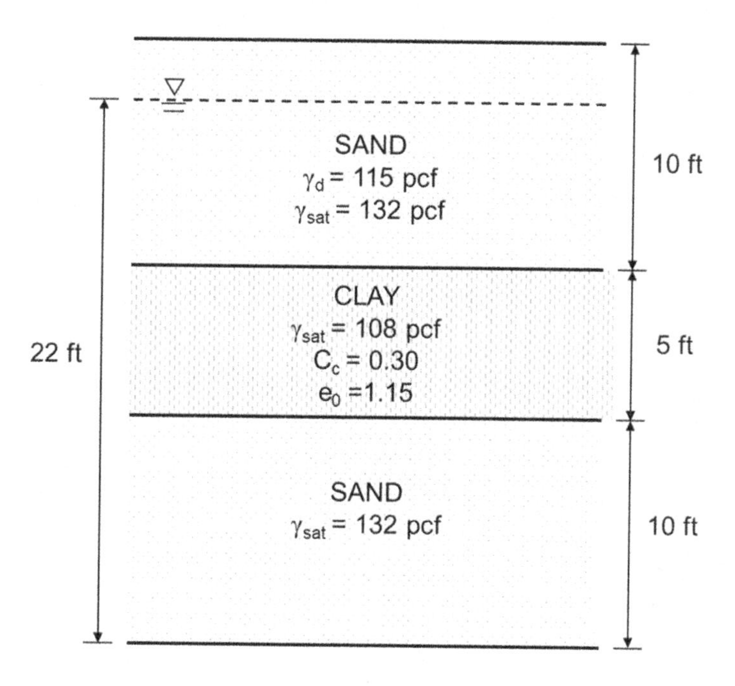

(A) 0.69 inches
(B) 1.15 inches
(C) 2.55 inches
(D) 2.75 inches

111. The footing shown below overlies soil with a unit weight equal to 130 lb/ft^3. The soil's friction angle is 34°, and the soil has zero cohesion. If the local regulator requires a factor of safety of 2.5 against bearing capacity failure, the allowable bearing capacity is most nearly:

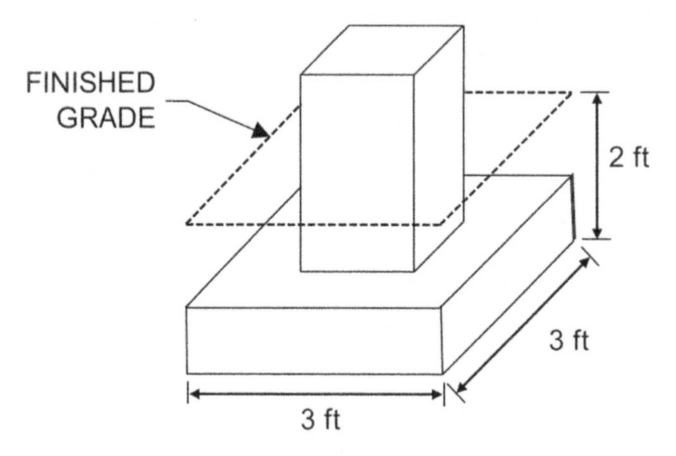

(A) 5 kips per square foot
(B) 7 kips per square foot
(C) 15 kips per square foot
(D) 18 kips per square foot

112. Earth Embankment A and earth Embankment B are equal, except that Embankment A is constructed of sandy gravels. Embankment B is constructed of silt. Which statement below is most correct?

(A) Embankment A has a higher factor of safety against slope failure, as sandy gravels generally have higher shear strength than silts.
(B) Embankment B has a higher factor of safety against slope failure, as sandy gravels generally have higher shear strength than silts.
(C) Embankment A has a lower factor of safety against slope failure, as sandy gravels never have higher shear strength than silts.
(D) Embankment B has a lower factor of safety against slope failure, as sandy gravels always have lower shear strength than silts.

113. After loading the column with square footing as shown below, the increase in vertical stress at Point "X" is most nearly:

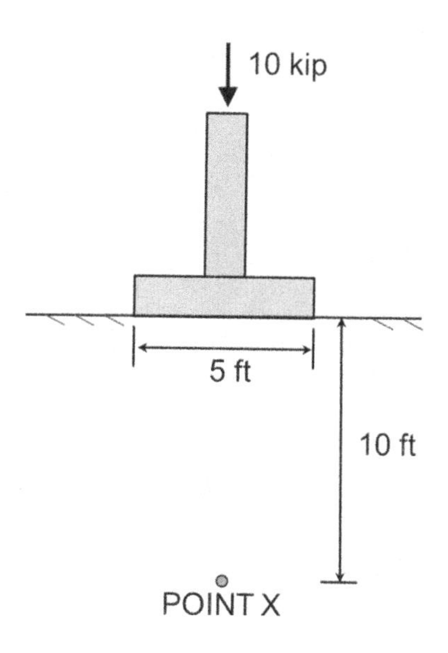

(A) 40 psf
(B) 60 psf
(C) 80 psf
(D) 100 psf

114. Given a maximum allowable bending stress of 4600 psi, the required depth (d) of the beam shown below is most nearly:

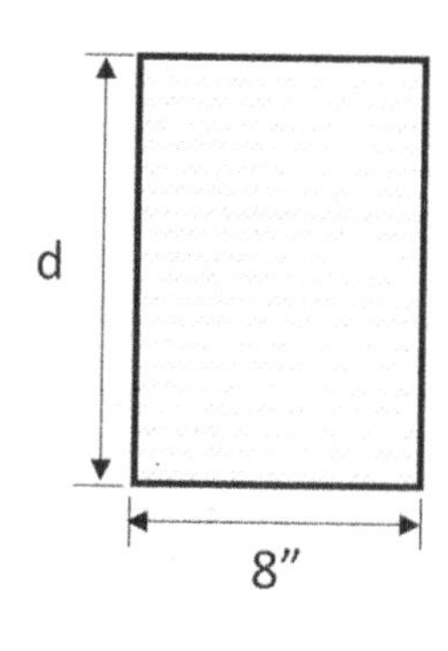

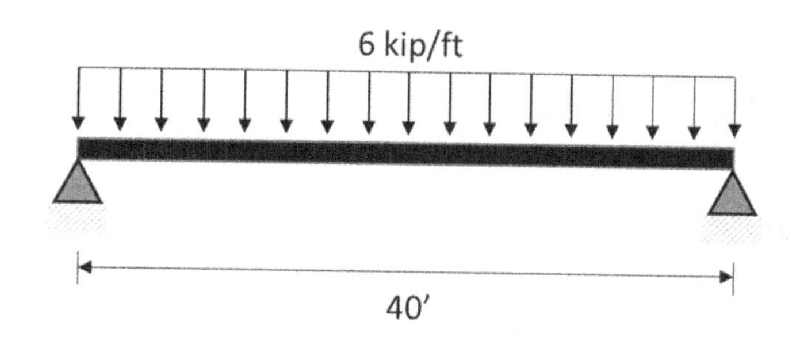

(A) 12 in
(B) 16 in
(C) 28 in
(D) 49 in

115. The number of zero force members in the truss shown below is most nearly:

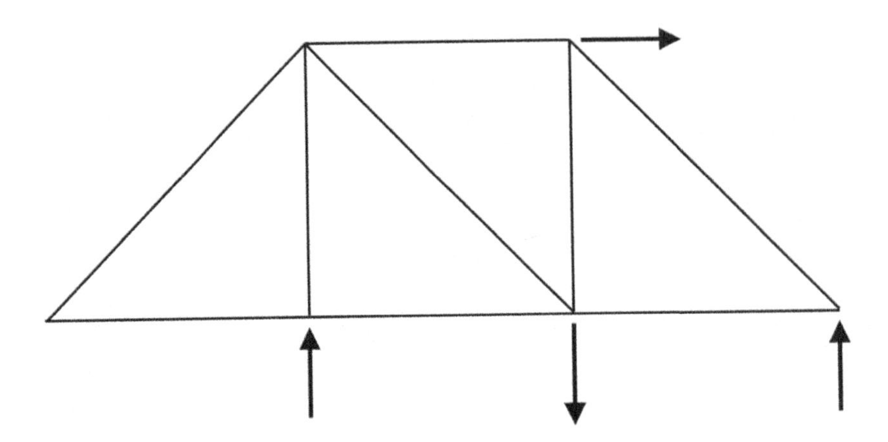

(A) 0
(B) 1
(C) 2
(D) 3

116. A 20-foot long steel cable [E=29(10^6) psi] with diameter equal to 3-inches is used to support a load of 7000 lbs. The elongation of the cable is most nearly:

(A) 0.008 inches
(B) 0.050 inches
(C) 0.800 inches
(D) 1.200 inches

117. Select the most correct statement below regarding ultimate strength design:

(A) Load factors for dead loads are generally higher than load factors for live loads.
(B) Load factors for live loads are generally higher than load factors for dead loads.
(C) Load factors for dead loads are generally equal to load factors for live loads.
(D) Load factors for live loads can be higher or lower than load factors for dead loads.

118. The maximum shear experienced in the beam shown below is most nearly: (take the vertically downwards direction as negative and the vertically upwards direction as positive)

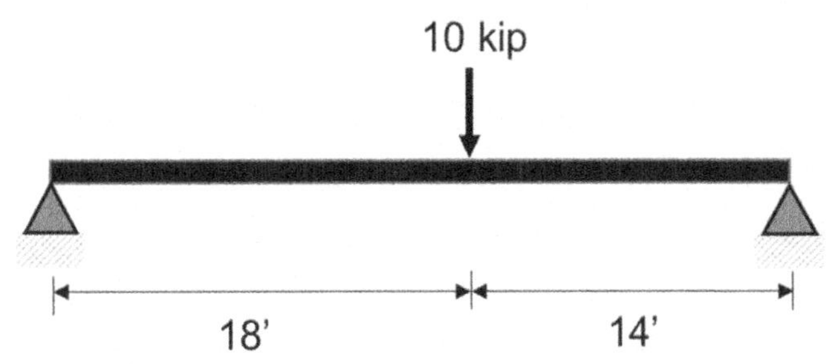

(A) -4.375 kip
(B) -5.625 kip
(C) 4.375 kip
(D) 5.625 kip

119. A concrete gravity retaining wall is shown below. The factor of safety against overturning at the toe is most nearly:

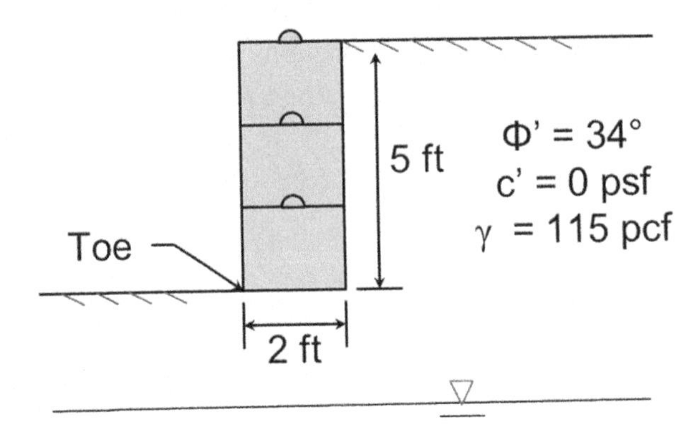

(A) 1.5
(B) 1.8
(C) 2.2
(D) 2.4

120. The following data is collected from a tipping cup style rain gauge during a storm with a duration equal to 4 hours.

Time (hr)	0.25	0.50	0.75	1.00	1.25	1.50	1.75	2.00	2.25	2.50	2.75	3.00	3.25	3.50	3.75	4.00
Precipitation (in)	0.05	0.05	0.10	0.10	0.10	0.20	0.20	0.40	0.30	0.20	0.10	0.05	0.05	0.05	0.05	0.05

The average rainfall intensity of the storm (in/hr) is most nearly:

(A) 0.51 in/hr
(B) 0.62 in/hr
(C) 1.45 in/hr
(D) 2.05 in/hr

121. A 1,200-foot length of 4-inch-diameter polyvinyl chloride (PVC) pipe (Hazen-Williams roughness C=135) conveys 85 gpm of water. The friction loss in the pipe is most nearly:

(A) 3.1 feet
(B) 6.2 feet
(C) 7.3 feet
(D) 9.2 feet

122. The maximum permitted velocity of water through a certain pipe is 10 ft/s. The design scenario is estimated to be the fire flow (1500 gpm) plus maximum day demand (800 gpm). The minimum required diameter of the pipe to convey the full flow rate at peak demand without violating the maximum velocity criteria is most nearly:

(A) 6 inches
(B) 8 inches
(C) 12 inches
(D) 16 inches

123. Examine the three cross-sections below. Which has the largest hydraulic radius for the given water depth?

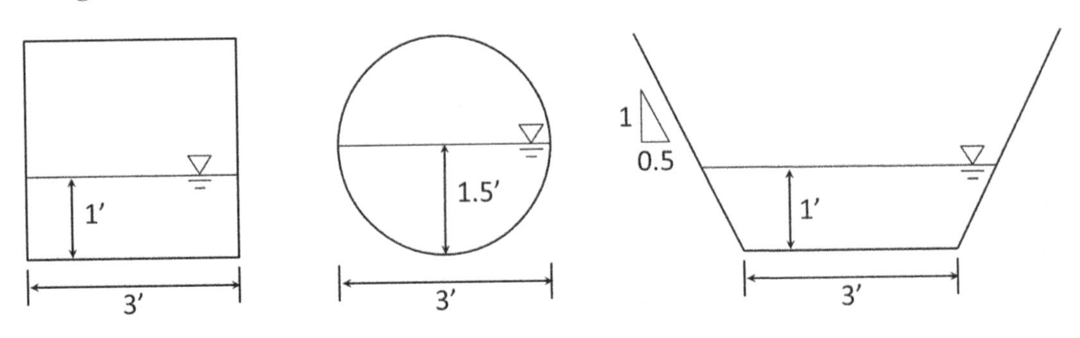

(A) The rectangular cross-section
(B) The circular cross-section
(C) The trapezoidal cross-section
(D) Each cross-section has the same hydraulic radius

124. Select the answer below that is most correct:

(A) In general, storms with shorter durations have higher rainfall intensities than storms with longer durations.
(B) In general, storms with longer durations have higher rainfall intensities than storms with shorter durations.
(C) In general, storms with longer durations have equal rainfall intensities to storms with shorter durations.
(D) In general, rainfall intensity is independent of storm duration.

125. A stormwater detention pond drains at a rate of 0.25 cfs. The pond's area is 0.10 acres, and there is 4 vertical feet of storage available in the pond. Assume vertical side slopes. If a long-duration storm results in an average runoff rate of 3 cfs into the pond, the amount of time before the pond reaches 1 foot of freeboard is most nearly:

(A) 1.3 hr
(B) 2.5 hr
(C) 3.4 hr
(D) 4.2 hr

126. Water flows from Point A to Point B, which are defined below:

Point A
Elevation: 100 feet
Pressure Head: 5 feet
Velocity Head: 1 foot

Point B
Elevation: 50 feet
Pressure Head: 40 feet

There is 10 feet of friction loss between Point A and Point B. The velocity at Point B is most nearly:

(A) 6.0 fps
(B) 14.5 fps
(C) 19.7 fps
(D) 32.1 fps

127. A vertical curve is shown below. The elevation at Point X along the curve is most nearly:

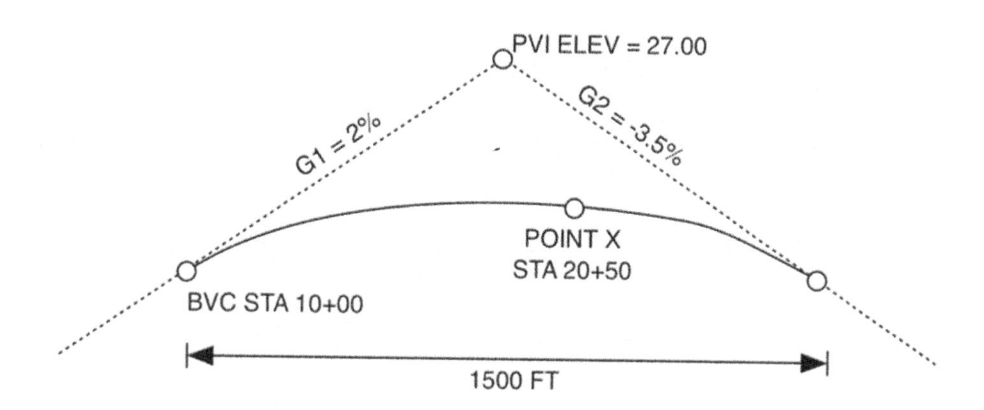

(A) 12.00
(B) 12.79
(C) 14.27
(D) 18.69

128. A traffic study is conducted on a rural highway, and the following data is collected for the eastbound lane:

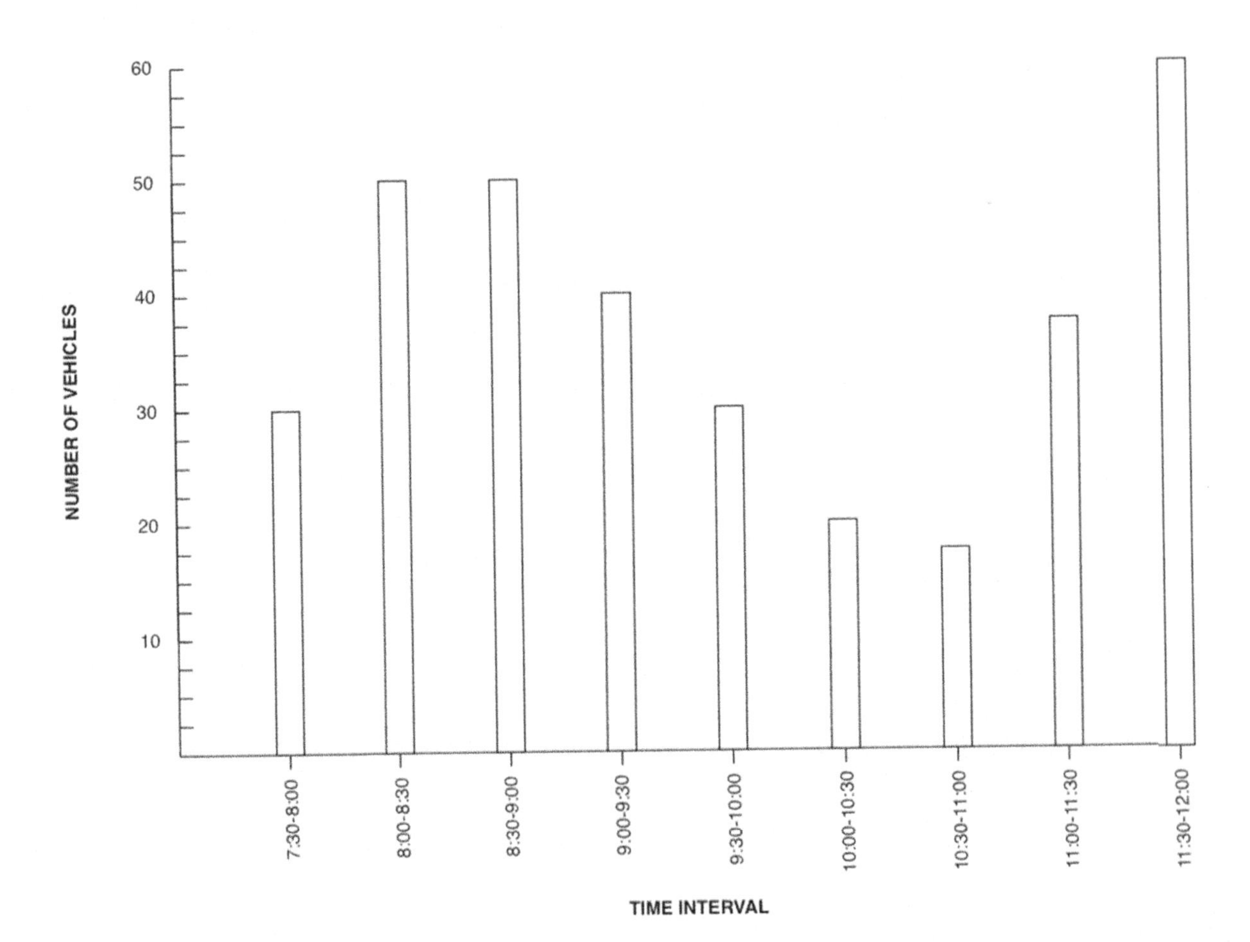

The peak hourly traffic volume is most nearly:

(A) 60 vehicles per hour
(B) 97 vehicles per hour
(C) 100 vehicles per hour
(D) 110 vehicles per hour

129. A horizontal curve has the properties listed below:

Item	Value
PT Station	13+50.00
PC Station	12+00.00
Radius	715 ft

The intersection angle of the curve is most nearly:

(A) 9 degrees
(B) 10 degrees
(C) 11 degrees
(D) 12 degrees

130. Which of the statements below regarding a soil's buoyant unit weight is the most correct:

(A) Buoyant unit weight is less than the soil's saturated unit weight by exactly the unit weight of water.
(B) Buoyant unit weight is the upward force of displaced water on the soil.
(C) Buoyant unit weight is used in the capillary zone directly above the groundwater surface.
(D) Buoyant unit weight equals the saturated unit weight when all voids are completely filled with water.

131. An earth embankment requires 10,000 CY of fill material compacted to 95% of maximum dry density as determined by modified Proctor. A borrow pit nearby has soil with the following characteristics:

Unit weight: 120 lb/ft^3
Water content: 8%
Maximum dry density (modified Proctor): 135 lb/ft^3

The volume of borrow material required to construct the embankment is most nearly:

(A) 9,833 CY
(B) 10,688 CY
(C) 11,543 CY
(D) 12,150 CY

132. A cast-in-place concrete retaining wall supports a critical highway, and the allowable duration of closure for a repair is limited. Which type of portland cement is the most appropriate for use in the concrete mix design?

(A) Type I
(B) Type II
(C) Type III
(D) Type IV

133. A permeability test is conducted on a cylindrical soil sample. Five cubic inches of water are percolated through the sample over 10 minutes, holding constant water levels as shown below. The sample's permeability is most nearly:

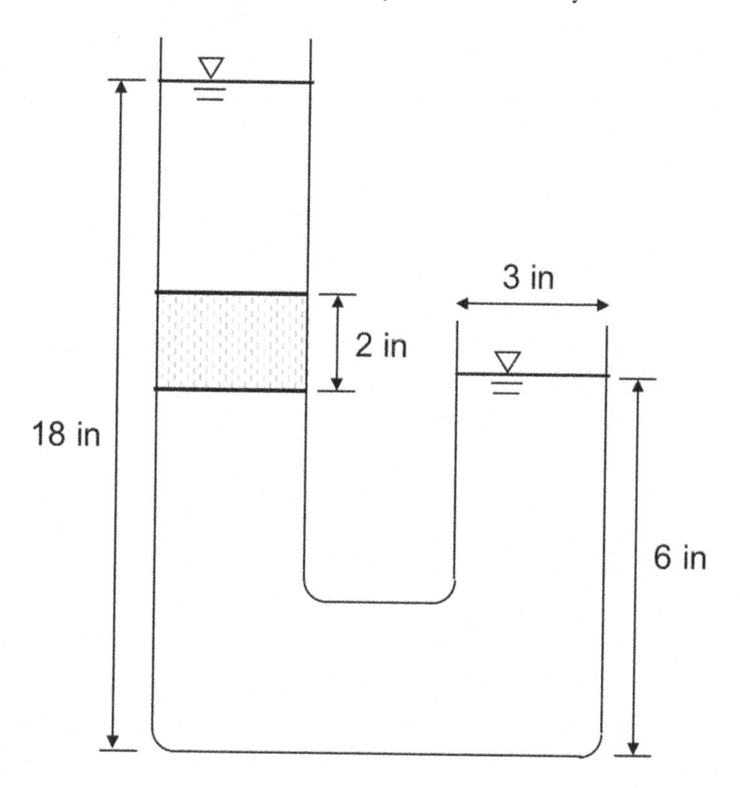

(A) 0.01 in/hr
(B) 0.18 in/hr
(C) 0.71 in/hr
(D) 1.06 in/hr

134. A soil has the following properties reported from a laboratory analysis:

Specific Gravity of Solids: 2.65
Porosity: 0.37

The buoyant unit weight of the soil is most nearly:

(A) 64.8 pcf
(B) 68.5 pcf
(C) 70.5 pcf
(D) 88.6 pcf

135. What is the primary difference between A36 and A992 (Grade 50) structural steel?

(A) A992 (Grade 50) steel has a higher density than A36 steel
(B) A992 (Grade 50) steel has a higher coefficient of thermal expansion than A36 steel
(C) A992 (Grade 50) steel has higher tensile strength than A36 steel
(D) A992 (Grade 50) steel is weathering steel, and A36 steel is not

136. A modified proctor test is performed on fill material, and the maximum dry density is reported as 135 pounds per cubic foot (pcf) at an optimum water content of 8%. A nuclear density gauge is being used in the field to determine if adequate material compaction is being achieved. The specifications call for each lift to be compacted to 95% of maximum dry density as determined by the modified proctor. What is the minimum density that must be reported by the nuclear density gauge to ensure compliance with the specification?

(A) 126.50 pcf
(B) 127.25 pcf
(C) 127.75 pcf
(D) 128.25 pcf

137. A contractor is required to fill an excavation with 150 compacted cubic yards of fill material. The fill material's swell has been estimated at 25%, and the material's shrinkage has been measured at 15%. The volume of fill required is most nearly:

(A) 173 loose cubic yards
(B) 176 loose cubic yards
(C) 188 loose cubic yards
(D) 221 loose cubic yards

138. The following cross-sectional areas of cut and fill are calculated from the construction of a roadway at 100-foot increments:

Station	Net Cross-Sectional Area of Cut/Fill
0+00	80 SF Cut
1+00	40 SF Cut
2+00	10 SF Fill
3+00	35 SF Fill

The net earthwork volume required to construct the road is most nearly:

(A) 75 CY Cut
(B) 165 CY Fill
(C) 165 CY Cut
(D) 195 CY Cut

139. What is the maximum permitted slope for an excavation in Type C soils, for excavations less than 20 feet deep?

(A) Vertical
(B) 0.75:1 (H:V)
(C) 1:1 (H:V)
(D) 1.5:1 (H:V)

140. In a temporary traffic control zone transition area, an upstream taper is created with channelization devices. The roadway has two-lanes in each direction, and the typical vehicle speed is 45 mph. The maximum recommended distance between channelization devices on an upstream taper is 1 foot per mph of vehicular speed. The maximum distance between channelization devices is most nearly:

(A) 1 foot
(B) 20 feet
(C) 45 feet
(D) 90 feet

END

CIVIL PE PRACTICE EXAM

BREADTH EXAM

VERSION B

SOLUTIONS

Problem	Solution	Problem	Solution
101	D	121	B
102	C	122	C
103	C	123	B
104	D	124	A
105	A	125	A
106	C	126	C
107	D	127	B
108	A	128	C
109	A	129	D
110	A	130	A
111	B	131	C
112	A	132	C
113	A	133	C
114	D	134	A
115	C	135	C
116	A	136	D
117	B	137	D
118	B	138	D
119	C	139	D
120	A	140	C

Solutions

101. Formwork Quantity Estimate
Primary Category: Project Planning
Secondary Category: Quantity Take-Off Methods

Approach: Calculate the area of each side of each wall shown in the figure and sum. Combine the walls of each height for ease of calculation.

Load Bearing Walls:

$$10\,ft\cdot(100\,ft+80\,ft)/_{side}\cdot 2\ {}^{sides}/_{wall}\cdot 2\,walls = 7{,}200\,ft^2$$

Non-Load Bearing Walls:

$$9\,ft\cdot[(80-30)\,ft+40ft)]/_{side}\cdot 2\,sides = 1{,}620\,ft^2$$

Sum:

$$7200\,ft^2+1620\,ft^2 = 8{,}820\,ft^2$$

102. Total Float Free Float Definition
Primary Category: Project Planning
Secondary Category: Project schedules

Approach: By definition, free float is the duration an activity can be delayed without delaying the early start date of immediately subsequent activities. Total float is the duration an activity can be delayed without violating a set schedule constraint or extending the project completion date.

103. Pavement Cost
Primary Category: Project Planning
Secondary Category: Cost Estimating

Approach: First, calculate the total quantity of asphalt or concrete required for each alternative. Next, multiply by the supplied unit costs and select the lower-cost option. For ease of calculation, consider calculating the plan-view area of the roadway first.

Area of Roadway

$$1\ mi \cdot {}^{5{,}280\ ft}/_{mile} \cdot \left(4\ lanes \cdot {}^{15\ ft}/_{lane} + 2\ shoulder \cdot {}^{8\ ft}/_{shoulder}\right) = 401{,}280\ ft^2$$

Asphalt Option Cost:

$$401{,}280\ ft^2 \cdot 8\ in \cdot {}^{1\ ft}/_{12\ in} \cdot {}^{145\ lb}/_{ft^3} \cdot {}^{1\ ton}/_{2000\ lb} \cdot {}^{\$100}/_{ton} = \$1{,}939{,}520$$

Concrete Option Cost:

$$401{,}280\ ft^2 \cdot 6\ in \cdot {}^{1\ ft}/_{12\ in} \cdot {}^{1\ yd^3}/_{27\ ft^3} \cdot {}^{\$270}/_{yd^3} = \$2{,}006{,}400$$

Asphalt is more cost-effective than concrete for this roadway.

104. Critical Path
Primary Category: Project Planning
Secondary Category: Project Schedules

Approach: The critical path is the path with the longest duration, because that path governs the duration of the entire project. To find the critical path, calculate the duration of each path to find which one is the longest. In this example, there are only three paths

Path A-B-C-F

The duration of A is 3 days and the duration of B is 5 days. C has a start-start lag relationship with B, so C begins 2-days after B starts. Since the duration of C is 2 days, C will be finished before B is finished. F begins once C is done and lasts for 2 days. Therefore, the duration of C does not contribute to the duration of the path. The total duration equals:

$$Duration\ Path\ ABCF = 3 + 5 + 2 = 10\ days$$

Path A-D-F

The duration of A is 3 days and the duration of D is 4 days. F has a finish-finish lag relationship with D, so F is completed 2 days after D is completed. Since the duration of F is 2 days, F will begin when D is completed and the finish-finish lag relationship will govern the duration of the path (i.e., the duration of F does not contribute to the duration of the path). The total duration equals:

$$Duration\ Path\ ADF = 3 + 4 + 2 = \ 9\ days$$

Path A-D-E-F

The duration of A is 3 days, the duration of D is 4 days, the duration of E is 4 days, and the duration of F is 2 days. Since all relationships are start-finish, each activity's duration contributes to the duration of the path.
The total duration equals:

$$Duration\ Path\ ADEF = 3 + 4 + 4 + 2 = \ 13\ days$$

105. Float
Primary Category: Project Planning
Secondary Category: Activity Identification and Scheduling

Approach: The way to solve this type of problem is to draw an activity-on-node network for the project with the supplied information. One can then calculate the early start date, late start date, early finish date, and late finish date for each activity. Float is calculated as late start (LS)-early start (ES) or the late finish (LF) – late start (LS).

For this problem though, it is not necessary to draw an activity-on-node network because only one path is described from the start of the project to the end of the project and that path is the critical path: place and compact base course, prepare formwork, pour concrete, and backfill behind the wall. While there is a finish-start lag described between the concrete pour and the placement of backfill (10 days for the concrete to cure), this is not float. Float is described as the amount of time that an activity can be delayed without delaying the overall schedule. From inspection, a delay in any of the described activities will extend the end date of the project. There is zero float for items on the critical path.

106. Excavation Depth
Primary Category: Means and Methods
Secondary Category: Construction Methods

Approach: Since the project has not yet begun, finished grade has not yet been established. This means the excavation depth is the distance from existing grade to the required subgrade elevation, as follows:

$$Depth = 120.85 - (110.65 - 2 - 0.5) = 12.7\,ft$$

107. Long Reach Boom
Primary Category: Means and Methods
Secondary Category: Construction Methods

Approach: This problem can be solved with geometry. First, check which arm length governs. Since Arm A cannot deflect past the horizontal, Arm B must be able to reach the bottom of the excavation. Additionally, if Arm B is long enough to reach the bottom of the excavation only, Arm A must be long enough to reach horizontally from the excavator over the cut slope of the excavation. Given the relationship between the length of Arm A and Arm B, the total length can be found once the governing arm length is calculated.

Minimum Length of Arm B:

$$L_B = 18\,ft + 10\,ft = 28\,ft$$

and:

$$L_B = 1.25 \cdot L_A, therefore\ L_A = \frac{L_B}{1.25} = \frac{28}{1.25} = 22.4\,ft$$

Minimum Length of Arm A:

$$L_A = \ 10\,ft + 18\,ft \cdot 0.25\frac{ft}{ft} = 14.5\,ft$$

$$L_B = 1.25 \cdot 14.5 = 18.13\,ft$$

Therefore the length of Arm B governs, and the total length of the long-reach boom is:

$$Total\ Length = 28\,ft + 22.4\,ft = 50.4\,ft$$

108. Active/Passive Earth Pressure
Primary Category: Soil Mechanics
Secondary Category: Lateral earth pressure

Approach: By definition, active earth pressures are mobilized when a retaining wall moves away from the backfill. Conversely, passive earth pressures are mobilized when a retaining wall moves into the backfill. Therefore, Resultant "A" is active and Resultant "B" is passive when analyzing slipping and overturning.

109. Total Stress
Primary Category: Soil Mechanics
Secondary Category: Effective and total stresses

Approach: Calculate the total load on the column by summing the live load and the dead load, then divide by the allowable bearing capacity to determine the minimum required area of the footing.

Total load on column:

$$2.25\ kip + 6\ kip = 8.25\ kip$$

Minimum footing area:

$$\frac{8.25\ kip \cdot 1000\ {}^{lb}/_{kip}}{0.75\ {}^{ton}/_{ft^2} \cdot 2000\ {}^{lb}/_{ton}} = 5.5\ ft^2$$

Minimum square length:

$$\sqrt{5.5\ ft^2} = 2'\ 4.1"$$

Select the next whole inch: 2' 5"

110. Clay Consolidation
Primary Category: Soil Mechanics
Secondary Category: Soil Consolidation

Approach: A normally consolidated clay is one which is currently experiencing its highest vertical stress. Primary consolidation of a normally consolidated clay is caused by the release of water from the soil voids, which is induced by a change in vertical stress. The soil consolidates and settlement occurs as water drains from the voids. The long term primary settlement can be calculated from the equation:

$$S = \frac{C_c}{1+e_0} H_0 \log_{10} \frac{p'_f}{p'_0}$$

First calculate the original effective vertical stress at the midpoint of the clay layer:

$$\begin{aligned} p'_0 &= (10\,ft + 5\,ft + 10\,ft - 22\,ft) \cdot 115\,{}^{lb}/_{ft^3} + (22\,ft - 5\,ft - 10\,ft) \\ &\cdot \left(132\,{}^{lb}/_{ft^3} - 62.4\,{}^{lb}/_{ft^3}\right) + 2.5\,ft \cdot \left(108\,{}^{lb}/_{ft^3} - 62.4\,{}^{lb}/_{ft^3}\right) \\ &= 946.2\,psf \end{aligned}$$

Next, solve directly for the primary consolidation with the soil parameters given:

$$S = \frac{0.3 \cdot 5\,ft \cdot \log_{10} \frac{946.2\,psf + 200\,psf}{946.2\,psf}}{1 + 1.15} = 0.058\,ft = 0.69\,in$$

111. Bearing Capacity
Primary Category: Soil Mechanics
Secondary Category: Bearing Capacity

Approach: The ultimate bearing capacity for a concentrically loaded square footing is:

$$q_{ult} = cN_c s_c + qN_q s_q + 0.5\gamma B_f N_\gamma s_\gamma$$

The first term does not apply because cohesion is zero. Note that:

$$q = q_{appl} + \gamma_a D_f$$

Again, the first term is zero because there is no surface surcharge. The shape correction factors are given as follows:

$$s_q = 1 + \left(\frac{B_f}{L_f}\right)\tan\phi = 1 + \left(\frac{3}{3}\right)\tan 34 = 1.67$$

$$s_\gamma = 1 - 0.4\left(\frac{B_f}{L_f}\right) = 1 - 0.4\left(\frac{3}{3}\right) = 0.6$$

The bearing capacity factors are published as a function of the friction angle, N_q=29.4 and N_γ=41.1.

$$q_{ult} = (130\,pcf)(2\,ft)(29.4)(1.67) + 0.5(130\,pcf)(3\,ft)(41.1)(0.6) \approx 17{,}575\,psf$$

Divide by the factor of safety to obtain the allowable bearing capacity:

$$q_{all} = \frac{17{,}575}{2.5} \approx 7{,}000\,psf \text{ or } {\sim}7\,ksf$$

112. Shear Strength
Primary Category: Soil Mechanics
Secondary Category: Slope Stability

Approach: Generally, sands and gravels (non-cohesive soils) have high shear strengths (i.e., friction angles) than silts and clays (cohesive/plastic soils). It must be noted that this is not universally true and depends somewhat on the clay mineralogy of fine-grained soils, but it is a generally accepted rule of thumb in practice. If two embankments are otherwise equal, the embankment constructed from the soil with a higher friction angle will have a higher factor of safety against slope failure. The factor of safety against slope failure is calculated by performing a moment equilibrium analysis, and balancing the forces supporting the embankment against the forces causing the embankment to fail. Shear strength is a significant force that supports the embankment and contributes significantly to the factor of safety.

113. Boussinesq Stress
Primary Category: Soil Mechanics
Secondary Category: Total and effective stress

Approach: First, find the stress that has been applied at the surface by dividing the column load by the area of the footing. Next, use a Boussinesq Stress Contour Chart (available in a variety of references) for square footings to identify the fraction of the applied stress that is applied at a depth of twice the length of the footing, in the center of the footing.

Applied stress:

$$q_0 = \frac{10\,kip \cdot 1000\,{}^{lb}/_{kip}}{5\,ft \cdot 5\,ft} = 400\,{}^{lb}/_{ft^2}$$

From the Boussinesq Stress Chart, the applied stress at depth = 2B in the center of a square footing is approximately $0.1q_0$.

Therefore, applied stress at Point "X" is

$$\Delta p = 0.1 \cdot 400\,{}^{lb}/_{ft^2} = 40\,{}^{lb}/_{ft^2}$$

114. Bending Stress in Beam
Primary Category: Structural Mechanics
Secondary Category: Bending (e.g., moments and stresses)

Approach: Set the bending stress in the beam equal to 4600 psi and solve for d to find the minimum allowable depth.

Bending stress in a beam is given by $\sigma = \frac{Mc}{I}$. In this case M is taken as the maximum moment in the beam, which for a uniformly loaded simply supported beam is given as $M = \frac{wL^2}{8}$ (and occurs at the center of the span). The variable c is the vertical distance from the neutral axis of the beam cross-section to the edge of the cross-section. The moment of inertia, I, is calculated for a rectangular section about the centroid as $I = \frac{bh^3}{12}$. Therefore:

$$M = \frac{wL^2}{8} = \frac{6\,{}^{kip}/_{ft} \cdot 1000\,{}^{lb}/_{kip} \cdot (40\,ft)^2\,(12\,{}^{in}/_{ft})}{8} = 14{,}400{,}000\,in \cdot lb$$

$$c = \frac{d}{2}\,in$$

$$I = \frac{bh^3}{12} = \frac{8 \cdot (d\,in)^3}{12}\,in^4$$

$$\sigma = \frac{Mc}{I} = 4{,}600\,psi = \frac{(14{,}400{,}000\,in \cdot lb)\,(\frac{d}{2}\,in)}{\frac{8 \cdot (d\,in)^3}{12}\,in^4} =$$

Solve for d

$d > 48.5$ in

115. Zero Force Members
Primary Category: Structural Mechanics
Secondary Category: Trusses

Approach: A zero force member occurs where a third member is joined to two collinear members and where only two non-collinear members join at an apex, if no external loads are applied at these locations. By inspection, only the two left-most members can be zero force members.

116. Elasticity
Primary Category: Structural Mechanics
Secondary Category: Axial

Approach: Recall that the Modulus of Elasticity is a ratio of the stress to strain in a material, meaning it directly relates the strain that will be imposed by a given stress. Second, recall that strain is the change in length over the length for an axial member. First solve for the stress in the cable imposed by the load, then solve for the strain induced in the cable, and finally solve for the change in length.

$$\sigma = \frac{7000\ lb}{\pi \cdot (\frac{3\ in}{2})^2} = 990.3\ {}^{lb}/_{in^2}$$

$$\varepsilon = \frac{\sigma}{E} = \frac{990.3\ {}^{lb}/_{in^2}}{29 \cdot 10^6\ {}^{lb}/_{in^2}} = 0.000034$$

$$\frac{\Delta L}{L} = 0.000034, \qquad \Delta L = 20\ ft \cdot 12\ {}^{in}/_{ft} \cdot 0.000034 = 0.008\ in$$

117. LRFD
Primary Category: Structural Mechanics
Secondary Category: Dead and Live Loads

Approach: In ultimate strength design, also known as load and resistance factor design (LRFD), load factors are applied to service loads to calculate the ultimate required strength of a member. Service loads are typically separated into live and dead loads, with live loads consisting of transient loads and dead loads consisting of permanent loads. Service loads are then factored by load factors, such as 1.2 or 1.4. In general, live loads have larger load factors. This is intuitive because live loads require a larger factor of safety. There is more uncertainty involved with live loads compared to dead loads.

118. Shear
Primary Category: Structural Mechanics
Secondary Category: Shear

Approach: The maximum shear in a simply supported beam with an intermediate point load can be found by solving for the reaction forces. The shear diagram can then be drawn if needed, but it has been generalized and is available from a variety of resources. Essentially, the right reaction will impose a positive shear (upwards) from zero, then the shear remains constant moving to the right until the point load is reached, where it will decrease by the magnitude of the point load (and become negative), then the shear remains constant again moving to the right until the left reaction where a positive shear is imposed that is exactly equal to the negative shear (causing the shear diagram to resolve at zero). Since the point load is closer to the right reaction, the area between the point load and the right reaction will experience the maximum shear (and it will be negative and equal to the right reaction).

The right reaction for a simple beam with intermediate point load is given by $R_r = \frac{Pa}{L}$ where a is the distance between the left reaction and the load.

$$Maximum\ Shear = -R_r = \frac{-Pa}{L} = \frac{-10\ kip * 18\ ft}{18\ ft + 14\ ft} = -5.625\ kip$$

119. Retaining Wall - Overturning
Primary Category: Structural Mechanics
Secondary Category: Retaining Walls

Approach: The factor of safety against overturning for a retaining wall is found by performing a moment balance around the toe. The factor of safety (FS) is calculated as $FS = \frac{\sum Resisting\ Moments}{\sum Overturning\ Moments}$. The resisting moment in this problem is caused by the weight of the wall. The overturning moment is caused by the resultant force from the active lateral earth pressure from the backfill behind the wall. Groundwater and cohesion can be ignored.

First, solve for the active lateral earth pressure resultant using Rankine Theory:

Calculate the coefficient of active earth pressure:

$$K_a = \tan^2\left(45 - \frac{\phi}{2}\right) = \tan^2\left(45 - \frac{34}{2}\right) = 0.283$$

Calculate the resultant magnitude from the backfill height and unit weight (per unit width of wall):

$$R_a = \frac{K_a \gamma z^2}{2} = 0.5 \cdot 0.283 \cdot 115\ {}^{lb}/_{ft^3} \cdot (5\ ft)^2 = 407\ lb$$

The resultant is located 2/3 of the distance down from the top of the backfill or 1/3 of the distance up from the toe, so the distance between the resultant and the toe is $5\ ft \cdot \frac{1}{3} = 1.67\ ft$

Therefore, the overturning moment about the toe is:

$$Overturning\ Moment = 407\ lb \cdot 1.67\ ft = 680\ ft \cdot lb$$

The resisting force is the weight of the concrete wall (per unit width of wall):

$$Weight\ Wall = 150\ {}^{lb}/_{ft^3} \cdot 5\ ft\ \cdot 2\ ft = 1{,}500\ lb$$

The weight of the wall acts at the midpoint of the wall, so the resisting moment is:

$$Resisting\ Moment = 1{,}500\ lb\ \cdot 1\ ft = 1{,}500\ ft \cdot lb$$

Therefore, the factor of safety against overturning can be calculated as:

$$FS = \frac{\sum Resisting\ Moments}{\sum Overturning\ Moments} = \frac{1{,}500\ ft \cdot lb}{680\ ft \cdot lb} = 2.2$$

120. Storm Intensity
Primary Category: Hydraulics and Hydrology
Secondary Category: Storm Characteristics

Approach: By definition, rainfall intensity is a measure of the amount of precipitation per hour and can be defined as an instantaneous intensity (which will vary over the storm duration), peak instantaneous intensity, or average intensity. Since we are asked for the average intensity, we must sum all rainfall measured over the storm and divide by the storm duration.

Average intensity

$$\frac{(7 \cdot 0.05 + 4 \cdot 0.1 + 3 \cdot 0.2 + 0.3 + 0.4)\ in}{4\ hr} = 0.51\ {}^{in}/_{hr}$$

121. Friction Loss
Primary Category: Hydraulics and Hydrology
Secondary Category: Pressure Conduit

Approach: The following equation can be used:

$$h_f = \frac{4.73L}{C^{1.852}D^{4.87}}Q^{1.852}$$

$$h_f = \frac{4.73(1{,}200\ ft)}{135^{1.852}\left(4\ in/_{12\ in/ft}\right)^{4.87}}\left(85\ gpm \cdot 1\frac{gpm}{448.8}cfs\right)^{1.852} \approx 6.2\ ft$$

122. Continuity
Primary Category: Hydraulics and Hydrology
Secondary Category: Energy and/or Continuity Equation

Approach: One of the two governing physical laws for closed-conduit hydraulics (the other being the energy equation) is the continuity equation (Q=VA). With a given, constant flow rate, fluid must flow faster in smaller pipes and flow slower in larger pipes. A simplified equation derived from Q=VA for the units of inches (diameter) and gpm (flow), fps (velocity) is $V = \frac{0.408Q}{d^2}$. Thus, we set V to the maximum permitted velocity, plug in the flow rate, and solve for d.

$$Q = 1500 + 800 = 2300\ gpm$$

$$V = 10\ fps = \frac{0.408 \cdot 2300\ gpm}{d^2}$$

Thus, d = 9.7 inches. Select the next largest available pipe size.

123. Hydraulic Radius
Primary Category: Hydraulics and Hydrology
Secondary Category: Open Channel Flow

Approach: Hydraulic radius is defined as $R_H = \frac{A}{P_W}$ where A is the area of flow and P_W is the wetted perimeter. Hydraulic radius is a measure of the hydraulic efficiency of a channel, it is the ratio of how much flow passes (area) per unit flow subject to friction loss from contact with the channel (wetted perimeter). Calculate the hydraulic radius for each cross-section.

Rectangular channel:

$$R_H = \frac{A}{P_W} = \frac{3\ ft \cdot 1\ ft}{3ft + 1\ ft + 1ft} = 0.6\ ft$$

Circular Channel:

$$R_H = \frac{A}{P_W} = \frac{\frac{\pi \cdot (\frac{3\,ft}{2})^2}{2}}{\pi \cdot \frac{3\,ft}{2}} = 0.75\,ft$$

Trapezoidal Channel:

$$R_H = \frac{A}{P_W} = \frac{3\,ft \cdot 1\,ft + 2 \cdot (\frac{1}{2} \cdot .5\,ft \cdot 1ft)}{3\,ft + 2 \cdot \sqrt{1^2 + 0.5^2}} = 0.67\,ft$$

124. Intensity-Duration-Frequency
Primary Category: Hydraulics and Hydrology
Secondary Category: Storm Characteristics

Approach: As evidenced by any typical storm intensity-duration-frequency curve, intensity increases as duration decreases. The relationship between the two parameters is inversely exponential. This is intuitive, as the rainfall during short thunderstorms is usually more "intense" (more precipitation falls per unit time) than rainfall during longer, more average storms.

125. Detention Sizing
Primary Category: Hydraulics and Hydrology
Secondary Category: Detention/Retention Ponds

Approach: This is essentially a mass balance problem. First calculate the volume of storage available, then calculate the net flowrate of volume into the pond. Finally, divide the storage by the flowrate to find the time until the given conditions is established (1 foot of freeboard). Freeboard is the distance between the top of the storage and the top of the water surface.

Volume Storage:

$$V = 0.10\,ac \cdot 43{,}560\,{}^{ft^2}/_{ac} * (4\,ft - 1\,ft) = 13{,}068\,ft^3$$

Net Flowrate in:

$$Q = 3\,{}^{ft^3}/_{s} - 0.25\,{}^{ft^3}/_{s} = 2.75\,{}^{ft^3}/_{s}$$

Time for storage to be filled at net flowrate:

$$13{,}068\,ft^3 \cdot \frac{1\,s}{2.75\,ft^3} \cdot \frac{1\,min}{60\,s} \cdot \frac{1\,hr}{60\,min} = 1.32\,hr$$

126. Energy Equation
Primary Category: Hydraulics and Hydrology
Secondary Category: Energy Equation

Approach: Use the energy equation to solve for the velocity head (and hence the velocity) at Point B.

$$\frac{P_A}{\gamma} + \frac{v_A^{\,2}}{2g} + z_A = \frac{P_B}{\gamma} + \frac{v_B^{\,2}}{2g} + z_B + h_l$$

$$5\,ft + 1\,ft + 100\,ft = 40\,ft + \frac{v_B^{\,2}}{2g} + 50\,ft + 10\,ft$$

$$\frac{v_B^{\,2}}{2g} = 6\,ft, \;\; v_B = 19.7\,{}^{ft}/_{s}$$

127. Vertical Curve
Primary Category: Geometrics
Secondary Category: Basic vertical curve elements
Approach: A vertical curve is completely described by the following equation:

$$Curve\ elevation = Y_{BVC} + g_1 x + x^2 \frac{g_2 - g_1}{2L}$$

First, calculate the elevation of the BVC given g_1, the PVI elevation, and L. Vertical curves are symmetrical over the PVI, so the distance from the PVI to the BVC is L/2.

$$Y_{BVC} = 27.00 - \frac{1500}{2} \cdot 0.02 = 12.00$$

$$Curve\ elevation = 12 + .02(1{,}050) + (1{,}050)^2 \frac{-.035 - .02}{2(1{,}500)} \approx 12.79$$

128. Peak Hourly Volume
Primary Category: Geometrics
Secondary Category: Traffic Volume

Approach: The peak hourly traffic volume is the maximum number of vehicles counted in a one-hour interval. The one-hour interval with the most vehicles is the interval from 8:00-9:00.

From 8:00-9:00 the total number of vehicles = 50+50 = 100 vehicles per hour

129. Horizontal Curve
Primary Category: Geometrics
Secondary Category: Circular Curves

Approach: Since we are given the station of the PT and the PC, we are given the length of the curve: $L = STA_{PT} - STA_{PC}$=150 ft. We can solve for the intersection angle using the radius and length as follows:

$$L = \frac{R\Delta\pi}{180}$$

$$150 = \frac{(715)\Delta\pi}{180} \quad thus \quad \Delta \approx 12\ degrees$$

130. Buoyant Weight Definition
Primary Category: Materials
Secondary Category: Soil properties (e.g., strength, permeability, compressibility, phase relationships)

Approach: By definition, the buoyant unit weight (also referred to as the submerged unit weight or effective unit weight) of a soil is equal to the unit weight of water subtracted from the submerged unit weight of the soil. Buoyant unit weight is used to calculate effective stress, which accounts for the reactions between individual soil particles and disregards the effects of water in the voids (i.e., water has no shear strength and doesn't contribute significantly to the support).

131. Borrow Quantity
Primary Category: Materials
Secondary Category: Soil properties (e.g., strength, permeability, compressibility, phase relationships)

Approach: Since there are two different densities at play (the in-situ density and the maximum dry density), we must find the mass (weight) of solids required from the borrow material. Next, use the given water content and wet density of the borrow material to calculate the dry density (because we must compare the weight of soil solids, not water). Use the weight of solids required in the embankment and the borrow material's dry density to find the total volume of borrow material needed.

Weight of soil solids needed:

$$10{,}000\ yd^3 \cdot 27\ {}^{ft^3}\!/_{yd^3} \cdot 0.95 \cdot 135\ {}^{lb}\!/_{ft^3} = 34{,}627{,}500\ lb$$

Dry density of borrow material:

$$\rho_d = \frac{\rho_{sat}}{1+w} = \frac{120\ {}^{lb}\!/_{ft^3}}{1+0.08} = 111.11\ {}^{lb}\!/_{ft^3}$$

Volume of borrow material required:

$$34{,}627{,}500\ lb \cdot {}^{1\ ft^3}\!/_{111.11\ lb} \cdot {}^{1\ yd^3}\!/_{27\ ft^3} = 11{,}543\ yd^3$$

132. Cement Type
Primary Category: Materials
Secondary Category: Concrete

Approach: Generally, Type I cement is "normal", Type II cement is "modified" (lower heat generation), Type III cement is "high early strength", and Type IV is "low heat" (lowest heat generation). Since this project requires a short cure time, Type III is the most appropriate. It should be noted that these are not the only differences between the different cement types. Another important performance characteristic that varies between the different types of cement is sulfate resistance, although that is not a factor in in this problem.

133. Permeameter
Primary Category: Materials
Secondary Category: Soil Properties

Approach: The two classic laboratory tests for soil permeability are the falling head and constant head permeameter tests. A falling head test measures the drop in liquid through the permeameter over time and relates that value (along with the geometry of the permeameter) to the sample's permeability. A constant head test measures the volume of water used to maintain a constant head over the sample, and relates that value (along with the geometry of the permeameter) to the permeability of the sample. This problem involves a constant head test. The equation for a constant head permeameter test is $K = \frac{QL}{tAh}$. Thus, the permeability is calculated as follows:

$$h = 18 - 6 = 12\ in$$

$$A = \pi \cdot \left(\frac{3}{2}\right)^2 = 7.07\ in^2$$

$$K = \frac{5\ in^3 \cdot 2\ in}{12\ in \cdot 7.07\ in^2 \cdot 10\ min \cdot {}^{1\ hr}/_{60\ min}} = 0.71\ in/hr$$

134. Phase Relationships
Primary Category: Materials
Secondary Category: Soil Properties

Approach: Phase relationship problems will always give you all of the knowns required to solve for an unknown. In this case, we are asked to solve for buoyant unit weight. Buoyant unit weight can be expressed in terms of the soil solid's specific gravity and the soil's void ratio as $\gamma_b = (\frac{G+e}{1+e} - 1)\gamma_w$. In addition, the porosity (given) can be directly related to the void ratio as $e = \frac{n}{1-n}$. Therefore, we can input the given parameters and solve directly for the buoyant unit weight:

$$e = \frac{n}{1-n} = \frac{0.37}{1-0.37} = 0.59$$

$$\gamma_b = \left(\frac{G+e}{1+e} - 1\right)\gamma_w = \left(\frac{2.65+0.59}{1+0.59} - 1\right) 62.4\ {}^{lb}/_{ft^3} = 64.8\ {}^{lb}/_{ft^3}$$

135. Structural Steel
Primary Category: Materials
Secondary Category: Structural Steel

Approach: By definition, A992 (Grade 50) steel is rated to a tensile yield strength of 50 ksi, while A36 steel is rated to a tensile yield strength of 36 ksi.

136. Compaction
Primary Category: Materials
Secondary Category: Compaction

Approach: In order to comply with the density specification requiring the material be compacted to 95% of the maximum dry density as determined by the modified proctor, the material must be compacted to the density below:

$$0.95 \cdot 135 \, {}^{lb}/_{ft^3} = 128.25 \, {}^{lb}/_{ft^3}$$

137. Swell and Shrinkage
Primary Category: Site Development
Secondary Category: Excavation and Embankment

Approach: The three possible volumes of a soil in construction earthwork are as follows: bank cubic yards, loose cubic yards, and compacted cubic yards. They are related by the following:

$$Loose\ CY = \frac{100 + \%\ swell}{100} \cdot Bank\ CY$$

$$Compacted\ CY = \frac{100 - \%\ shrinkage}{100} \cdot Bank\ CY$$

Therefore

$$Loose\ CY = \frac{100 + \%\ swell}{100 - \%\ shrinkage} \cdot Compacted\ CY$$

$$Loose\ CY = \frac{100 + 25}{100 - 15} \cdot 150\ CY = 221\ CY$$

138. End-Area Method
Primary Category: Site Development
Secondary Category: Excavation and Embankment

Approach: Given various cross sectional areas with the intent to calculate a volume, utilize the average end area method. This method involves averaging two areas at either end of a distance, then multiplying the average area by the length in question to determine a volume. For ease of calculation, create a table as follows:

Given Data:

Station	Net Cross-Sectional Area Cut/Fill
0+00	80 SF Cut
1+00	40 SF Cut
2+00	10 SF Fill
3+00	35 SF Fill

Tabulations:

Station Interval	Average End Area (SF)	Volume (CY)
0+00 to 1+00	(-80-40)/2=-60 Cut	(-60)(100)/27=-222 CY Cut
1+00 to 2+00	(-40+10)/2=-15 Cut	(-15)(100)/27=-56 CY Cut
2+00 to 3+00	(10+35)/2=22.5 Fill	(22.5)(100)/27=83 CY Fill
	TOTAL	-222-56+83=-195 CY Cut

139. Excavation Safety
Primary Category: Site Development
Secondary Category: Safety

Approach: Per OSHA 1926.652 B, the maximum allowable slope for a Type C soil is 1.5:1 (H:V).

140. Traffic Control
Primary Category: Site Development
Secondary Category: Safety

Approach: The maximum distance between channelization devices on an upstream taper shall be 1 ft per mph of vehicle speed. Therefore, the maximum distance between channelization devices is:

$$1\ {}^{ft}/_{mph} \cdot 45\ mph = 45\ ft$$

Made in the USA
Monee, IL
16 August 2022